PTEROSAURS
THE FLYING REPTILES

BY S. CHRISTOPHER BENNETT, PH.D.

Illustrated by Brian Franczak and Bob Walters
with additional artwork by Charles A. Parker

PREHISTORIC LIFE
FRANKLIN WATTS
NEW YORK / CHICAGO / LONDON / TORONTO / SYDNEY

Photographs courtesy of the author

Library of Congress Cataloging-in-Publication Data

Bennett, S. Christopher.
Pterosaurs : the flying reptiles / by S. Christopher Bennett;
illustrated by Brian Franczak and Bob Walters
p.   cm.
Includes bibliographical references and index.
ISBN 0-531-11181-4
1. Pterosauria—Juvenile literature.   [1. Pterosaurs.
2. Dinosaurs.] I. Franczak, Brian, ill.   II. Title.
QE862.P7B45   1994
567.9′7—dc20   93-29845
                                               CIP AC

Contents

This skeleton of *Pterodactylus* was the first
pterosaur fossil ever found.

1
The Pterosaurs Appear

Early in the **Mesozoic** Era, which we often call the Age of Dinosaurs, a strange group of animals appeared on Earth. These were the pterosaurs—the word means winged lizards—and they were the first vertebrates, or animals with backbones, that had the ability to fly. The first pterosaurs were small and not particularly good fliers, but during the next 150 million years they evolved to become excellent fliers. Many different shapes and sizes appeared, and some grew to gigantic size and were the largest flying animals that ever lived.

We know about pterosaurs and other animals that lived long ago because some of them were buried in mud at the bottom of rivers and seas. As the mud dried and became stone, their bones were preserved. These buried remains of animals are called **fossils**.

The first fossil of a pterosaur was found in the mid-1700s, long before dinosaur remains were discovered, in a region that is now southern Germany. The fossil was an amazingly complete skeleton of a small pterosaur, perfectly preserved on a slab of fine limestone.

Pterosaurs in the Mesozoic Era

The fourth finger on each hand of the fossil was very long, and people were puzzled because they had never seen such an animal before. At first they thought it was some sort of swimming animal with long flippers, like a penguin. Later the French biologist Georges Cuvier showed that its skeleton was like the skeletons of reptiles. He also said that the long finger had supported a wing, and that the creature could fly. Cuvier declared it was a flying reptile, and he called it a *Pterodactyle*, based on the Latin for wing finger.

Since then fossils of pterosaurs have been found all over the world, on all continents. However, most pterosaur fossils come from just a few places, which have been named for the type of rock in which the fossils were found. These are the Solnhofen Limestone of Germany, the Cambridge Greensand near Cambridge, England, the Niobrara Chalk of Kansas in the United States, and the Santana Formation of northeastern Brazil.

Flying Vertebrates

Birds and bats are the only types of flying vertebrates that are living now. Birds, bats, and pterosaurs are not related. They are quite different from one another, and each developed the ability to fly independently. The most noticeable differences are in their wings. In all three—birds, bats, and pterosaurs—the forelimb has been modified into a wing with long slender arm bones. In bats the hand is large, with four long slender fingers and a short thumb. The arm and fingers support and stretch a leathery membrane that forms the surface of the wing. Only the thumb is free for climbing and grasping. In birds, most of the bones of the hand and fingers are **fused** together. Long, strong feathers that form the surface of the wing are

8

This drawing shows how the wings of birds
(top), bats (center), and pterosaurs (bottom)
differ in structure.

attached to the arm and hand; the arm and almost all of the body are covered by small feathers that insulate and streamline the bird.

In pterosaurs, the first three fingers of the hand were short and had large claws at the end, but the fourth finger was very large and long. Most of the bones of the wrist were fused together and a new bone, the pteroid, was present. The arm and fourth finger supported a tough membrane that stretched from the finger to the side of the body and the hind leg. This membrane formed the wing surface and was very different from the bats' leathery wing membrane. The outer part had fibers, that helped stiffen it and gave it the right shape for flying. The pteroid bone controlled the smaller membrane in front of the arm, and the first three small fingers were free for grasping and climbing. Pterosaur wings were as different from those of birds and bats as bird and bat wings are different from each other.

Two Types of Pterosaurs

There were two main types of pterosaurs. The first pterosaurs were the **rhamphorhynchoids**, named for the **genus** *Rhamphorhynchus* (beak snout). They had long tails, and the metacarpals, the bones that form the palm of the hand, were short. The second group of pterosaurs evolved from a rhamphorhynchoid late in the **Jurassic** Period (see Time Scale) of the Mesozoic Era. These were the pterodactyloids, named for the genus *Pterodactylus* (wing finger). They had short tails and long metacarpals. We know of many more different kinds of pterodactyloids than rhamphorhynchoids, and some of them evolved to become enormous.

Open Questions

Fossils of pterosaurs are very rare. Bones of pterosaurs were thin, allowing the skeletons to be light so the animals could fly. But thin bones are very fragile, and so they often broke up before they could be fossilized. Pterosaur fossils are also rare because many of the animals lived in environments where their bodies or bones did not get buried in mud or sand, and so were not fossilized.

Because pterosaur fossils are very rare, there were probably many different species we don't know about.

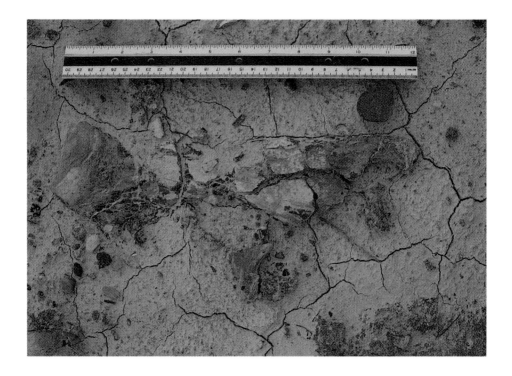

This crushed *Pteranodon* humerus
(upper-arm bone) was found weathering
out of the rock of the Niobrara Chalk in
western Kansas.

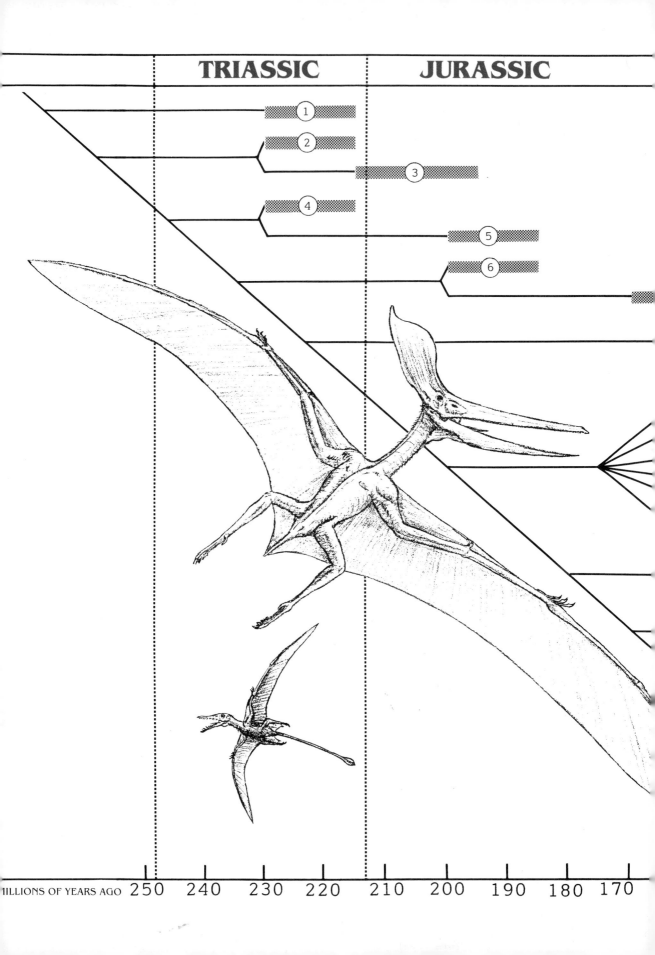

TRIASSIC   JURASSIC

① ② ③ ④ ⑤ ⑥

MILLIONS OF YEARS AGO 250 240 230 220 210 200 190 180 170

CRETACEOUS

RHAMPHORHYNCHOIDS

1 *Preondactylus*
2 *Peteinosaurus*
3 *Dimorphodon*
4 *Eudimorphodon*
5 *Campylognathoides*
6 *Dorygnathus*
7 *Rhamphorhynchus*
8 *Sordes*
9 *Scaphognathus*

PTERODACTYLOIDS

10 *Pterodactylus*
11 *Germanodactylus*
12 *Gnathosaurus*
13 *Huanhepterus*
14 *Ctenochasma*
15 *Pterodaustro*
16 *Nyctosaurus*
17 *Dsungaripterus*
18 *Ornithocheirus*
19 *Quetzalcoatlus*
20 *Ornithodesmus*
21 *Criorhynchus*
22 *Anhanguera*
23 *Tropeognathus*
24 *Pteranodon*

The Pterosaur Age
The numbered bars represent the pterosaur
genera (check the legend for genus names)
and are placed to show the time period
when each genus lived. The connecting lines
show the relationships among the genera.

7 8 9 10 11 12 13 14 15 16 17 18 19 20 21 22 23 24

160 150 140 130 120 110 100 90 80 70 60

There is also a lot that we don't yet know about pterosaurs because fossils tell us only a little about them. Soft parts of the body almost never are fossilized, so all we usually have are skeletons.

We are not certain what other reptiles were closely related to the pterosaurs. Some scientists think pterosaurs and dinosaurs evolved from the same ancestor; and if this is so, pterosaurs were closely related to dinosaurs. Other scientists believe that pterosaurs and dinosaurs evolved from different types of animals, and so were not closely related. However, fossils do show us that even the earliest known pterosaur can be recognized as a pterosaur, with wings and other special features that only pterosaurs had. We need to find new fossils of pterosaurs and other ancient animals to solve the puzzle of pterosaur origins.

2
Early Rhampho-rhynchoids: The First Pterosaurs

Pterosaurs first appeared sometime in the **Triassic** Period, 248 to 213 million years ago. The earliest known pterosaurs are *Preondactylus* (finger from Preone), *Peteinosaurus* (winged reptile), and *Eudimorphodon* (true two-shapes-of-teeth). Their fossil remains were all found in northern Italy. These animals were small, with wingspans of only about 2 to 3.3 feet (60 cm to 1 m). They were different enough from one another that scientists classify them in different **families** of pterosaurs. The very first pterosaur—the ancestor of all pterosaurs—must have appeared some time earlier.

Preondactylus seems to be the most primitive of the known pterosaurs because its wings were small and short for the size of its body, and its hind limbs were rather large. This suggests that *Preondactylus* was not as good a flier as later pterosaurs. *Preondactylus* and *Peteinosaurus* both had simple cone-shaped teeth, and may have eaten insects or small fish. *Eudimorphodon*'s teeth had more than one point and this pterosaur probably ate insects, because multipointed teeth are not good for catching and eating fish.

15

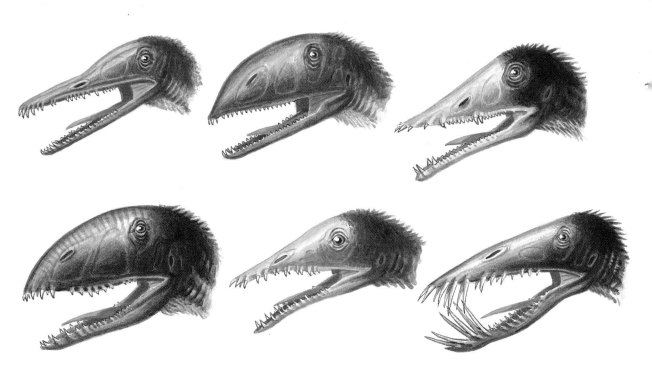

Left to right (top row): *Preondactylus,
Peteinosaurus, Eudimorphodon;* (bottom row)
Dimorphodon, Campylognathoides, Dorygnathus

Other early pterosaurs include *Dimorphodon* (two shapes of teeth), *Campylognathoides* (curved jaw), and *Dorygnathus* (spear jaw) that lived in the early part of the Jurassic Period (213 to 144 million years ago). Their fossils have been found in rocks that formed at the bottom of shallow seas near the coastlines of today's England and Germany.

Dimorphodon was of medium size, with a 4.5-foot (1.4-m) wingspan. Its large head was fairly long and narrow. It had many small, simple cone-shaped teeth in the back of its jaws, and larger cone-shaped teeth in the front. These teeth would have been good for catching fish, and so we believe *Dimorphodon* ate fish.

16

Campylognathoides was larger than *Dimorphodon* and had a wingspan of 5.7 feet (1.75 m). Its long, low skull held simple cone-shaped teeth, with the teeth in the front of the jaw only a little larger than those in the back. *Dorygnathus* was smaller than *Dimorphodon*, and had a wingspan of 3.3 feet (1 m). The very large and long teeth in the front of its jaw would have been perfect for catching and holding fish.

These pterosaurs probably flew much the way modern birds do, rapidly beating their long, slender wings in a smooth and graceful flight. They would not fly in the jerky, flitting manner of bats. However, if you saw one in flight, its long tail and large head would immediately signal that it was not a bird, but something very different.

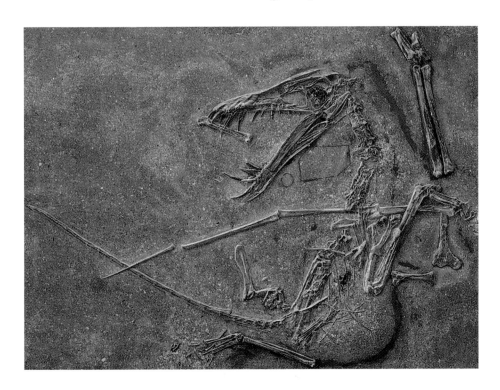

A complete skeleton of *Dorygnathus*
is preserved on a slab of dark shale,
a form of rock.

There are two theories of the origin of flight in pterosaurs. The "from the ground up" theory (below) suggests that running animals began to leap into the air to catch food and to escape predators, and eventually evolved flight. The "from the trees down" theory (above) suggests that tree-living animals began to glide from tree to tree, and eventually evolved flight.

Two Theories of Flight

Some scientists think that pterosaurs developed the ability to fly by running and jumping from the ground. They suggest that the ancestors of pterosaurs, the pre-pterosaurs, could run quite fast and that they fed on insects. They might have chased insects and jumped into the air trying to catch them. Perhaps they also ran and jumped into the air to escape from predators. Once in the air, they might have been able to control the direction of their jump, or increase the distance they could cover, by swinging their arms. Then, through many generations, wing membranes evolved. These increased the effect of arm swinging and so increased the height of the jump. Now the pre-pterosaurs could glide for some distance. Eventually the animals started to flap their wings and fly. This ''from the ground up'' theory of flight is favored by scientists who think pterosaurs were related to dinosaurs. If pterosaurs and dinosaurs were related, the earliest pterosaurs—like early dinosaurs—probably ran along the ground on their hind legs.

Other scientists think that pre-pterosaurs developed flight by leaping and gliding from trees. According to this "from the trees down" theory, the pre-pterosaurs lived in the trees, probably feeding on insects. Their long forelegs and hind legs enabled them to jump from one tree to another. Leaping from tree to tree would be easier than climbing down one tree, running across the ground, and climbing up the next tree. It was also safer, because by jumping from tree to tree the animals could avoid predators on the ground. Through many generations the wing membrane developed. The animals' control increased and they were able to glide. This increased the distance they could travel. As the animals became able to flap their wings, they could cover even more distance, and so powered flight was developed.

The theory that pterosaurs evolved from small animals that jumped from tree to tree seems reasonable. Today some animals live in trees, feed on insects or fruits, and glide from tree to tree. These include flying

Perhaps *Dimorphodon* ran along the ground
on two legs, like a bird.

Flying squirrels and the lizard *Draco* are
two tree-living animals that glide from tree
to tree on "wings" of skin that extend out
from their bodies.

squirrels, flying lemurs, and the gliding lizard *Draco*. The
animals glide from tree to tree on "wings" formed by
flaps of skin that extend horizontally from the sides of the
body. In flying squirrels and flying lemurs the skin
stretches from the forelimbs to the hind limbs. In the
gliding lizard the skin is held out by long, flexible ribs. So
pterosaurs probably evolved from small, long-legged liz-
ardlike animals that had a flap of skin extending from the
sides of their bodies and stretching from the forelegs to
the hind legs. The fourth finger became longer and larger
to increase the size of the wing so that the animals could
fly better.

Whichever way pterosaurs developed flight—from
the ground up or from the trees down—it is likely that the

pre-pterosaurs fed on insects. Pterosaurs probably began to feed on fish only after they had developed flight. They might have perched in trees, or flown over the water, and then dove down at small fish at the surface of the water. If this is so, then those pterosaurs like *Eudimorphodon*, that seem to have fed on insects, may be more primitive than the fish eaters.

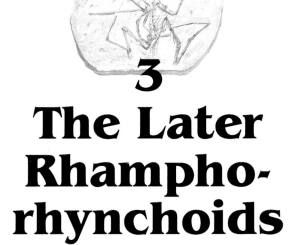

3
The Later Rhampho-rhynchoids

The rhamphorhynchoids continued to flourish through the late Jurassic Period. *Rhamphorhynchus, Scaphognathus* (scoop jaw), and *Sordes* (evil spirit) are the best-known late Jurassic rhamphorhynchoids.

Rhamphorhynchus was a medium-sized pterosaur with a wingspan of up to 5.9 feet (1.8 m) and a body about the size of a duck. Its long slender teeth were

Rhamphorhynchus, Scaphognathus,
and *Sordes*

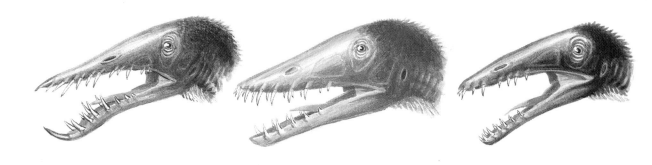

pointed forward and were perfect for grasping fish. The upper and lower jaws had no teeth at the tips, but tapered to slightly hooked points that were covered with horny beaks. Like other rhamphorhynchoids, *Rhamphorhynchus* had a long tail. A diamond-shaped vertical vane at the end of the tail might have been used as a rudder to steer its flight. The bones of the tail were interlocked, and this stiffened the tail. A stiff tail would have made it easier to use the rudder vane for steering.

 Scaphognathus was smaller than *Rhamphorhynchus*. It had a 3-foot (1-m) wingspan and a slightly deeper skull, with large teeth that pointed up and down rather than forward. However, its large teeth suggest that it, too, might have fed on fish. *Sordes* was similar to *Scaphognathus*, although smaller still, with a 2-foot (60-cm) wingspan. We know about *Sordes* from just a few fossils from central Asia, but they are excellent and show impressions of the wing membrane and even what looks like fur covering the body. One specimen also seems to show the wing membrane attached down the hind legs to the ankles, and a membrane between the hind legs and the tail.

A large skull of *Rhamphorhynchus* displays forward-pointing teeth well suited for catching fish.

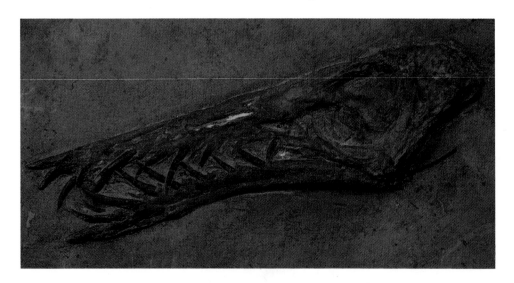

A complete skeleton of a small
Rhamphorhynchus, with a wingspan of only
12 inches and a vane at the end of the tail

A fossil wing of *Rhamphorhynchus*
preserves the flat membrane that
formed the wing surface.

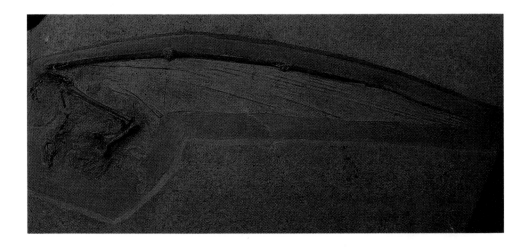

Hairy *Sordes*

Parke '95

A Fossil Record

We know a lot about *Rhamphorhynchus* because we have found large numbers of fossils. Many—found on slabs of limestone near Solnhofen, Germany—are even complete skeletons that were deposited in shallow lagoons behind reefs. *Rhamphorhynchus* probably flew out from shore and fed on fish in the lagoons. **Plankton** sometimes poisoned the fish and the pterosaurs that fed on them. Some pterosaurs died and fell to the bottom of the shallow lagoon. Since the fish life in the lagoon had been poisoned, there were no scavengers to eat the bodies of the pterosaurs, and so nothing disturbed the skeletons. The flesh decayed and the skeletons were buried by fine mud and eventually fossilized. Sometimes soft body parts were covered by the mud before they decayed and left impressions that we can see in the limestone. So some specimens of *Rhamphorhynchus* show the shape of the wing and beak. Sometimes the soft parts themselves were fossilized, but more often there is just an impression of the soft parts, showing where they were.

Fossilized impressions of the wing of *Rhamphorhynchus* show that it was long and narrow for most of its length. The wing attached to the hind limb and so would have been broad near the body, but narrower toward the tip. The impressions also show details of the wing membrane. Near the body it might have been able to stretch in all directions (like a bat's wing membrane), so that the wing could fold up. But the outer two-thirds of the wing had stiffening fibers that strengthened it and held it spread out.

Rhamphorhynchus fed on fish, and specimens have been found with masses of fish bones—a last meal—lying inside their rib cages. Males and females of *Rhamphorhynchus* differed slightly in proportions, and the

The fossilization of *Rhamphorhynchus*

A standing *Rhamphorhynchus*, ready to
spring into the air to seize a jumping fish

males had slightly larger heads. Fossils of all sizes of *Rhamphorhynchus* have been found. Young animals probably could fly or we would not find them as fossils. They probably left the nest early and did not need parental care as they grew to full size.

From all the fossils tell us, we can imagine a *Rhamphorhynchus* standing on the ground on its four legs. The stiff membranous wings are folded back over its furry body and long, scaly tail. It holds its head up on its long neck; its large, long teeth jut out of its mouth, and it has sharp, bright eyes. It moves quickly on the ground, stops to look around, watches for small fish jumping in the lagoon, and then quickly springs into the air to chase its prey.

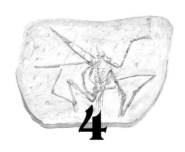

4
A New Type of Pterosaur: The Early Pterodactyloids

Even as the rhamphorhynchoids flourished, another group of pterosaurs appeared that, in time, replaced them—the pterodactyloids. The pterodactyloids were an advanced group of pterosaurs that evolved from a rhamphorhynchoid sometime in the Jurassic Period. The earliest pterodactyloids we have found date from the late Jurassic. Pterodactyloids had larger heads, longer necks, and shorter tails, than the rhamphorhynchoids, and the metacarpal bones that formed the palm of the hand were longer.

Pterodactylus is probably the best known genus of early pterodactyloid. There were a number of different **species**, ranging from small to middle-sized animals, with wingspans of from 4 to 7 feet (1.2 to 2.1 m). Most *Pterodactylus* specimens have been found in the Soln-hofen Limestone of Germany. Some are excellent, and

Two *Pterodactylus*, flying on their long, narrow wings

show impressions of the wing membrane, the flesh of the neck, and other soft parts of the body. Fish bones preserved inside some specimens' ribcages tell us *Pterodactylus* fed on small fish. Very small specimens, with mouse-size bodies, and wingspans of only 7 inches (17 cm), were also found. These were very young animals that may have been learning to fly when they died and were fossilized.

We have fossils of other early pterodactyloids. *Germanodactylus* (German finger) looked like *Pterodactylus* but had a low bony crest running along the top of its skull. *Gnathosaurus* (jaw reptile) and *Ctenochasma* (comb mouth) were also about the size of *Pterodactylus* with wingspans of 4 and 5.5 feet (1.2 and 1.6 m), respectively.

A skull of *Pterodactylus* shows the long, slender jaws with many small teeth.

Germanodactylus, Gnathosaurus, and
Ctenochasma

Both had long, narrow jaws with many slender curving teeth in each jaw. *Gnathosaurus* had about 130 teeth; those in the front of the jaws were quite long while those farther back were shorter. They must have made a perfect fish trap. *Ctenochasma* had longer teeth that were more closely spaced, and so had twice as many teeth as *Gnathosaurus*. The close spacing suggests that *Ctenochasma* fed on very small fish or crustaceans that it filtered out of the water. *Gnathosaurus* and *Ctenochasma*, like *Germanodactylus*, had bony crests on their heads. We do not know why they had crests.

Pterodaustro (southern wing) was an odd pterosaur from Argentina with a wingspan of 4 feet (1.3 m), and long, very slender upward-curving jaws. The upper jaw had very small teeth and the lower jaw had about a thousand long, slender teeth, or bristles, that must have been used to filter food from the water. It probably swished its jaws through the water to catch small animals, crustaceans, insects, or perhaps small plants. Although the teeth or bristles make *Pterodaustro* look similar to *Gnathosaurus* and *Ctenochasma*, we don't know if they were closely related.

Pterodaustro's thousands of slender teeth
in its lower jaw formed a sieve that filtered
tiny animals out of the water.

Huanhepterus (wing from Huanhe) was a mid-sized pterosaur from China with a wingspan of 5 feet (1.5 m), a long, slender head, and very long neck. The jaws were full of long, sharp teeth, and it probably fed on fish, like most other pterosaurs. Its head was slightly longer than its body, and its neck was half again as long, so *Huanhepterus* had a very long reach.

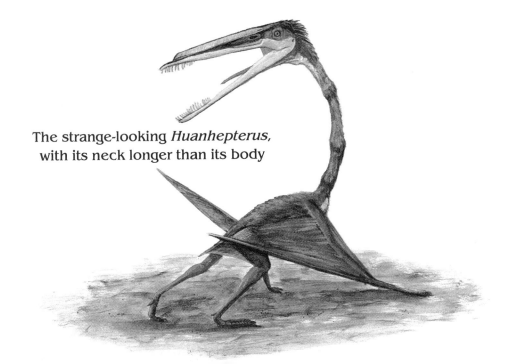

The strange-looking *Huanhepterus*,
with its neck longer than its body

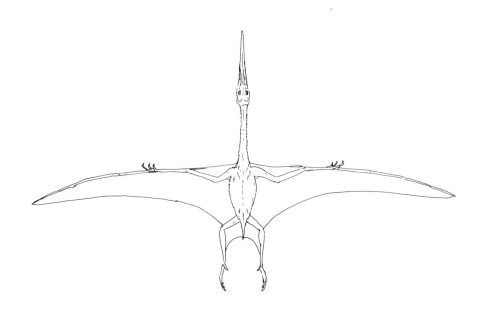

These drawings show the differences in
the length of the head, neck, metacarpals
(bones of the palm), and tail between
a pterodactyloid (top) and
a rhamphorhynchoid (bottom).

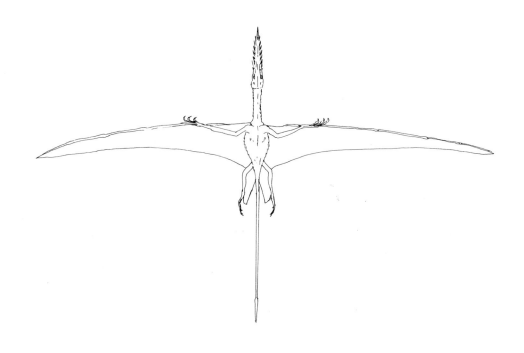

The Time of the Pterodactyloids

The pterodactyloids evolved from a rhamphorhynchoid some time in the Jurassic Period. Thus *Pterodactylus, Germanodactylus, Gnathosaurus,* and *Ctenochasma* lived side by side with *Rhamphorhynchus* and *Scaphognathus,* in and around the lagoons where the Solnhofen Limestone was deposited. Later in the Jurassic, the rhamphorhynchoids died out. Perhaps they could not compete with the pterodactyloids for food.

The pterodactyloids show more diversity than the rhamphorhynchoids. They ranged from small to middle sized, with strange crests on their heads, and lots of variation in body proportions and feeding methods. As we will see, the pterodactyloids continued to show varied living styles throughout the Mesozoic Era.

As pterodactyloids evolved, there were major changes in the skeleton. The most obvious changes were the pterodactyloids' short tail and long metacarpals that evolved from the long tail and short metacarpals of the rhamphorhynchoids. A short tail made the skeleton lighter, so the animal did not have to work as hard to fly. The short tail tells us, too, that the animal was an excellent flier: it did not need a long tail for balance and to control its flight. The pterodactyloids' longer metacarpals changed the wing proportions, making it longer and better suited for soaring flight. It must also have changed the way the animal walked on the ground.

We are not sure how pterosaurs moved about. Some scientists think pterosaurs walked on two legs, like birds or some dinosaurs; some think they walked on four legs; while others think they could only crawl. Scientists disagree because we don't know exactly how the forelimbs moved. If they could not be used for walking, the animals either had to walk on their hind legs or crawl on their

Rhamphorhynchoids may have crawled or
walked on all four legs (top and center).
Pterodactyloids with extremely long
forelimbs (bottom) may have walked
on just their hind legs.

bellies. However, we know the pre-pterosaurs could walk
quite well; they had to because they could not fly. Rhamphorhynchoids had four legs of about equal length, so
they could probably walk on all four legs, too. But some
pterodactyloids had forelimbs that were much longer
than their hind legs, and so they probably could not walk
on four legs.

5
The Heyday of the Pterodactyloids

The early part of the **Cretaceous** Period (144 to 65 million years ago) was the heyday of the pterodactyloids. They began to grow very large and there were many different genera. However, some are known from only a few fossils.

The fossils we have of early Cretaceous pterosaurs were found mostly in rocks in two places. The first is a large deposit of sands and muds near Cambridge, England, called the Cambridge Greensand. The Greensand contains many lumps of phosphate of lime that were mined for use as fertilizer. Some lumps contain fossils, and over 2,000 pterosaur fossils were found in them during the 1860s and 1870s. The fossils were mainly just pieces—of wing bones, jaws, and other odd parts of the skeleton—and so not enough of any animal was found to reconstruct what it looked like with certainty. Most of the fossils are from two genera, *Ornithocheirus* (bird hand) and *Criorhynchus* (ram snout). *Ornithocheirus* was of medium size with long slender jaws and a wingspan of 10 feet (3 m). *Criorhynchus* was larger, with a wingspan of 16 feet (5 m). It also had long jaws, but the upper and lower jaws had large rounded crests at the ends. The crests made the skull look very odd.

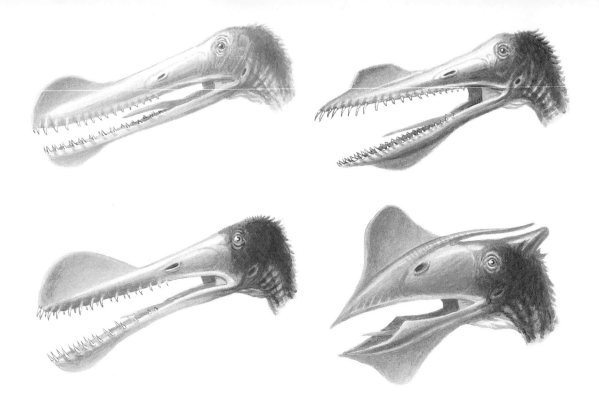

Left to right (top row): *Criorhynchus, Anhanguera;*
(bottom row) *Tropeognathus, Tapejara*

The second important source of early Cretaceous fossils is in northeastern Brazil. Here rocks called the Santana Formation contain concretions, which are masses of limestone that formed around animals and bones soon after they were buried in mud. The limestone prevented the bones from being crushed. When the limestone concretions are dissolved away with acid, perfectly preserved bones are freed. The bones are then studied and reconstructed, and we can see how the joints between bones fit together and how they work. This helps us to learn how the animals flew and walked.

The Santana Formation of Brazil is especially interesting because fossils of many sorts of pterosaurs have been found there, some of them similar to the fossils from the Cambridge Greensand. *Anhanguera* (old devil) seems very similar to *Ornithocheirus* from the Cam-

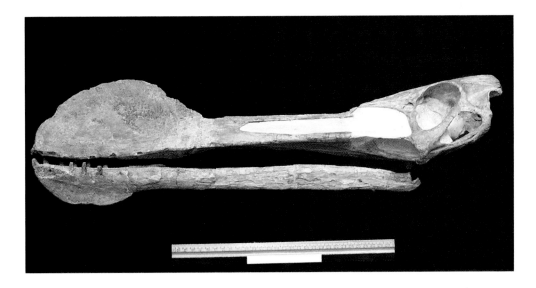

When a concretion was dissolved away,
this skull of *Tropeognathus* was revealed.

Two pterosaur neck vertebrae preserved
in a concretion, or mass of limestone, from
the Santana Formation in Brazil.

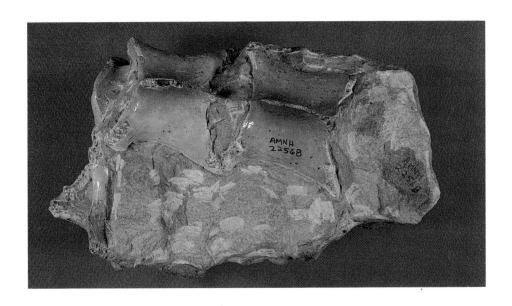

bridge Greensand, and *Tropeognathus* (keeled jaw) seems similar to *Criorhynchus*. Fossils of *Anhanguera* and *Tropeognathus* from Brazil are much better preserved than the Greensand fossils, and so tell much more about the animals. *Anhanguera* had small low crests at the ends of its jaws, and some scientists think that *Anhanguera* and *Tropeognathus* were actually females and males of the same sort of pterosaur.

Tapejara (old being) was a very strange looking pterosaur from the Santana Formation. It was a middle-sized pterosaur, with a short skull with a large, tall crest on top and no teeth. The jaws would not have been good for catching fish, so *Tapejara* probably ate fruits and seeds.

Aside from the Cambridge Greensand and the Santana Formation, early Cretaceous pterosaur remains have been found in other regions, too. *Dsungaripterus* (wing from Dsungaria) has been found in western China and central Asia. *Dsungaripterus* was a large pterosaur with a wingspan of 10 feet (3 m), a medium length neck, and a strange skull. It had a low crest on the top of its skull, and another small crest sticking out at the back. The front of the jaws had no teeth, but instead had curving beaks that tapered to points. The teeth in the middle of the jaws were slender, and at the back of the jaws they were short and cone-shaped. Almost all the teeth were worn flat on the top. This suggests that they were used for eating hard, rough food, probably clams and mussels. The curving beaks might have been used to dig up clams, which were then crushed between the back teeth. Perhaps *Dsungaripterus* walked along the beach at low tide, plucking clams and mussels from the mud, or perhaps it swam in the water and dove down to get shellfish, the way a duck dives for food.

Dsungaripterus and *Ornithodesmus*

Ornithodesmus (bird ribbon) was another unusual pterosaur that has been found in southern England. It was large, with a wingspan of 16 feet (5 m), and had long jaws with small triangular teeth just at the rounded tip of the jaws. The jaws would be good for biting out chunks of food, but do not seem suited for catching fish. We do not know what this animal fed on.

In some of these early Cretaceous pterosaurs, the bony crest on the upper arm bone, or humerus, where big wing muscles attach, had a peculiar shape. In other early Cretaceous pterosaurs, such as *Dsungaripterus*, the crest did not have the odd shape. The peculiar crest suggests that *Ornithodesmus, Anhanguera, Tropeognathus,* and *Criorhynchus* were closely related, and are in a family of pterosaurs called pteranodontids.

Pterosaurs Become Larger

Pterosaurs of the Cretaceous Period were generally much larger than the earlier pterosaurs of the Jurassic. Cretaceous pterosaurs probably were able to grow to large

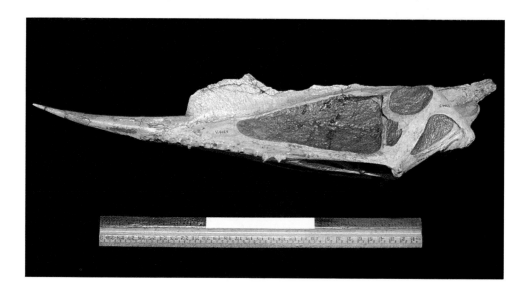

A skull of *Dsungaripterus*, with a pointed,
toothless beak, large teeth in the back of
the jaw, and odd crests on top

size because they evolved a number of special features.
Many of the bones in their spines fused and their shoulder blade (the scapula) had a joint with the spine. They
had hollow, air-filled bones, and some of them did not
have teeth. The fused spine and the joint with the scapula
made the animal's back and shoulders much stronger.
The back and shoulders could then hold out the huge
wings and withstand the strain put on the bones by the
large muscles that flapped the wings. Cretaceous
pterosaurs also evolved hollow bones. These made the
skeleton lighter, so the animals did not have to work as
hard to fly. Air-filled bones weigh less than bones filled
with fat or marrow, and can also be stronger. Another
special feature in some, but not all, early Cretaceous
pterodactyloids was the absence of teeth. Teeth are quite
heavy, so if a pterosaur could catch food without teeth, it
was better not to have them.

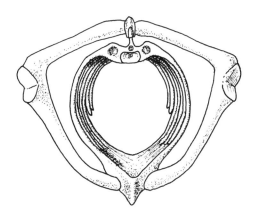

The typical shoulder girdle of
rhamphorhynchoids and some pterodactyloids,
seen from the front (left) and from the side (right).

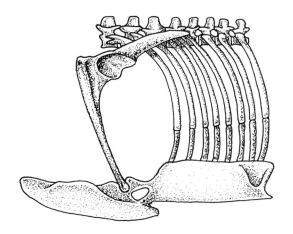
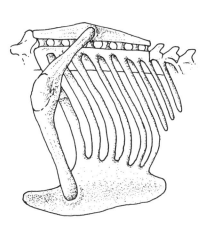

The typical shoulder girdle of large pterodactyloids,
seen from the front (left) and from the side (right).
The scapula, or shoulder blade, has
a joint with the backbone. This made the
shoulder girdle much stronger.

6
The Giant Pterodactyloids

Many late Cretaceous pterosaur fossils have been found, but only a few types are known. These include *Pteranodon* (wing without teeth), *Quetzalcoatlus* (named for the Aztec feathered snake god Quetzalcoatl), and *Nyctosaurus* (night reptile). *Pteranodon* is the best known, and many excellent specimens of it and of *Nyctosaurus* have been found in a deposit of rock in western Kansas called the Niobrara Chalk. The Chalk is made up of the shells of microscopic animals that lived and died in the middle of a seaway that connected the Arctic Ocean and Gulf of Mexico during much of the Cretaceous Period. For millions of years their shells sank to the bottom of the sea and became rock. The seaway was over 300 miles (500 km) wide, and was the home of many kinds of sharks and fish, large sea turtles, and large marine reptiles called mosasaurs and plesiosaurs.

Pteranodon was large, with a wingspan of up to 21 feet (6.25 m). It had long, tapering jaws with no teeth, and a large bony crest on the top of its skull. *Pteranodon* also had the peculiar crest on its humerus, like *Ornithodesmus* and *Anhanguera*, and so was related to them.

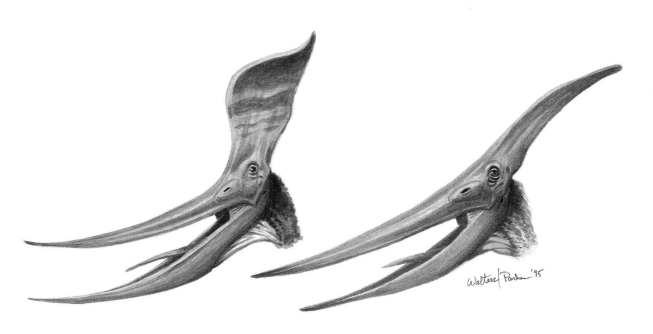

Pteranodon sternbergi and
Pteranodon longiceps

There were two species of *Pteranodon*. The earlier species was *Pteranodon sternbergi*, in which the large crest stuck up from the top of the skull. *Pteranodon sternbergi* evolved into the second species, *Pteranodon longiceps*, in which the crest was long and narrow and stuck out and back from the skull. Females were smaller than males and had smaller crests. *Pteranodon* had long, narrow wings that seem well-adapted to soaring, and it probably spent most of its time soaring over the sea, like the modern albatross. *Pteranodon* fed on fish, and sometimes rather large ones—up to a foot (30 cm) long. One odd feature of *Pteranodon* is that the upper jaw stuck out beyond the end of the lower jaw. You might think that the longer upper jaw made it harder to catch fish, but apparently *Pteranodon* managed.

Pteranodon soaring and fishing

Quetzalcoatlus

    Quetzalcoatlus was the largest known pterosaur, with a wingspan of 36 feet (11 m). It had long, slender, toothless jaws and a bony crest in the middle of its skull above the eyes. Its wings were shorter and broader than those of *Pteranodon*. The most unusual feature of *Quetzalcoatlus* was its very long neck. *Quetzalcoatlus* and other long-necked species from the late Cretaceous are called azhdarchids. They had very long necks and broad wings. The shape of their humerus suggests that they were more closely related to *Dsungaripterus* than to *Pteranodon* and *Anhanguera*.

    Another odd feature of *Quetzalcoatlus* and the other azhdarchids is that the outer wing bones were T-shaped (if you look at a cross-section) and had thick walls. Most other pterodactyloids had round or oval, thin-walled wing bones. A T-shaped bone weighs more than a round or oval bone, but is more resistant to damage from hitting the ground or from attacks by other animals. The hind

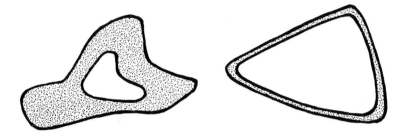

The wing bones of azhdarchids had thick
walls and were T-shaped in cross section
(left); thus they were less likely to break
than were the thin-walled wing bones (right)
of other pterodactyloids.

legs of *Quetzalcoatlus* were also bigger than those of
Pteranodon. From this evidence, it seems probable that
azhdarchids spent more time on the ground than did
other pterosaurs. *Quetzalcoatlus* may have been like a
giant heron or stork that walked along river and stream
banks, or waded in shallow water to feed on fish, frogs,
and other small animals.

Nyctosaurus was a small pterosaur with a wingspan of
up to 10 feet (3 m). *Nyctosaurus* showed a surprising
mixture of primitive and advanced features. It was primitive in that its shoulder blade, the scapula, did not have
the joint with the backbone that we see in other Cretaceous pterosaurs. This suggests that *Nyctosaurus* was a
side branch of pterosaur evolution. On the other hand,

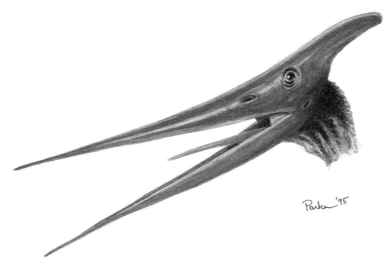

Nyctosaurus

Nyctosaurus seems advanced because it had toothless jaws and its wing finger had only three bones. All other pterosaurs had four bones in the wing finger. Fossils of *Nyctosaurus* have been found in the Niobrara Chalk of Kansas, and like *Pteranodon* it probably flew out over the Niobrara Sea and fed on fish.

Even Bigger and Better Pterosaurs

Pteranodon, Quetzalcoatlus, and *Nyctosaurus* all show that the skeletons of late Cretaceous pterosaurs were becoming more specialized for larger size, lighter weight, and better flying. They all had toothless jaws and the bones were getting even lighter. The very long neck bones of *Quetzalcoatlus* and other azhdarchids were thin-walled hollow tubes. Another change seen in these late Cretaceous pterosaurs is that the metacarpals grew

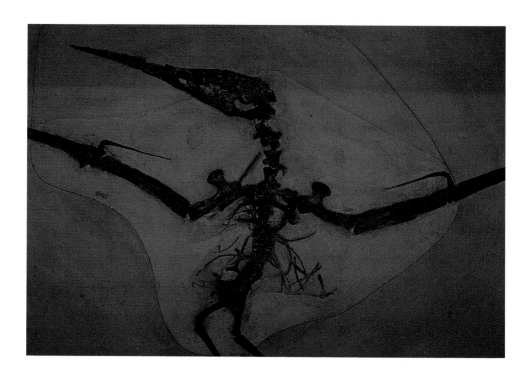

This skeleton of *Nyctosaurus* was found in
the Niobrara Formation of western Kansas.

very long: in *Nyctosaurus*, the metacarpals alone were longer than the entire hind limb. In all these animals, the metacarpals grew to be so long that there was no way these pterosaurs could have walked on four legs. Either they did not walk much, or they walked on just their two hind feet. They probably could stand on four legs, but walked on their hind legs, holding their heads high in the air.

These large pterodactyloids would have flown differently from the smaller pterosaurs. They would have flapped their wings when taking off, but with their huge wings, the flapping would have been slow. They might have had to run along the ground to build up speed for a take-off, but once in the air, they could have soared like

Large pterosaurs like *Pteranodon* probably
stood nearly upright and walked on their hind
legs (left); their long metacarpals would have
made walking on all four legs (right) difficult.

vultures, albatrosses, and other large modern birds.
Pteranodon had long slender wings like an albatross, and
probably soared gracefully over the water, steering by
making small adjustments to the position of the wings,
but without flapping its wings much. This is an efficient
way to fly because the animal uses the power of the wind
rather than its own muscle power. *Quetzalcoatlus* had
shorter, broader wings and probably soared over the hills
and plains of the ancient landscape, circling in currents of
rising air the way vultures do today.

A soaring *Quetzalcoatlus*

7
Extinction

The pterosaurs died out at the end of the Cretaceous, at the same time that the dinosaurs did. Before that, during the later part of the Cretaceous, the pterosaur genera seems to have decreased in number; but it may be that we have not yet found the fossils. If there was a real decrease, it may have been because the pterosaurs kept getting bigger. There are many more different kinds of small birds than big birds, and it would probably have been the same with pterosaurs. (In all animals there are many more species of small size than species of large size.) Or the pterosaurs may have had to compete for food with birds. At any rate, at the end of the Cretaceous, something caused enormous changes in the world, so that many types of animals became extinct. This may have been a change in climate that caused the world to become cooler, or a meteor or asteroid may have struck the Earth. Whatever caused the dinosaurs and many other animals to die, also killed the pterosaurs. All we can do is imagine how wonderful and beautiful they must have been as they flapped and soared through the skies.

The Pterosaur Genera: Pronunciation Guide

Anhanguera: ahn-hahn-GWER-a

Campylognathoides: camp-il-og-nath-OYD-eez

Criorhynchus: kry-o-RINK-us

Ctenochasma: teen-o-KAZ-ma

Dimorphodon: di-MORF-o-don

Dorygnathus: dor-eg-NATH-us

Dsungaripterus: tsung-ar-IP-ter-us

Eudimorphodon: YU-di-morf-o-don

Germanodactylus: gir-mahn-o-DAK-til-us

Gnathosaurus: nath-o-SOR-us

Huanhepterus: hwahn-HEP-ter-us

Nyctosaurus: nik-to-SOR-us

Ornithocheirus: or-nith-o-KY-rus

Ornithodesmus: or-nith-o-DEZ-mus

Peteinosaurus: pet-eye-no-SOR-us

Preondactylus: pre-on-DAK-til-us

Pteranodon: ter-AN-o-don

Pterodactylus: ter-o-DAK-til-us

Pterodaustro: ter-o-DOW-stro

Quetzalcoatlus: ket-zal-co-AHT-ul-us

Rhamphorhynchus: ram-fo-RINK-us

Scaphognathus: scaf-og-NATH-us

Sordes: SOR-deez

Tapejara: top-e-ZHAR-a

Tropeognathus: tro-pee-og-NATH-us

Glossary

(Individual pterosaurs discussed in the book are not defined in the glossary.)

Cretaceous (cree-TAY-shus): The last period of the Mesozoic Era. It spanned from 144 to 65 million years ago.

Family: A group of closely related species or kinds of animals or plants.

Fossil (FOS-sil): A buried trace of animals or plants that lived long ago.

Fuse: To melt and blend together. Two or more bones are said to be fused if they have grown together and become one larger bone.

Genus (GEE-nus): Plural—**genera**. A group of very closely related species or kinds of animals or plants.

Jurassic (jur-ASS-ik): The middle period of the Mesozoic Era. It spanned from 213 to 144 million years ago.

Mesozoic (mez-o-ZO-ik): The middle era of Earth's history. It spanned from 248 to 65 million years ago, and is often called the Age of Dinosaurs.

Plankton (PLANK-ton): Tiny plants and animals that live in water. Although some are poisonous, they are food for small animals that in turn are food for larger animals. When conditions are right some types may grow out of control or bloom. This may kill animals if they are poisonous, or the plankton may die and pollute the water, or possibly use up all the oxygen in the water.

Pterodactyloid (ter-o-DAK-til-oyd): The more advanced pterosaurs with short tails.

Rhamphorhynchoid (ram-fo-RINK-oyd): The early pterosaurs with long tails.

Species (SPEE-sheez): A single kind of animal or plant that can breed and reproduce. Each species has two names; a genus name and a species name. For example, *Pteranodon longiceps* is a species. *Pteranodon* is the genus name, and *longiceps* is the species name.

Triassic (try-ASS-ik): The first period of the Mesozoic Era. It spanned from 248 to 213 million years ago.

For Further Reading

Desmond, Adrian. *The Hot-Blooded Dinosaurs*. New York: Dial Press/James Wade, 1976.

Wellnhofer, Peter. *The Illustrated Encyclopedia of Pterosaurs*. New York: Crescent Books, 1991.

Index

ABOUT THE AUTHOR

S. Christopher Bennett is assistant professor in the Department of Systematics and Ecology at the University of Kansas, and is the author of many research papers and articles about pterosaurs. He is a graduate of Tulane University and earned an M.S. in biology at Yale University and a M.Phil. and Ph.D. in systematics and ecology from the University of Kansas.

ABOUT THE ILLUSTRATORS

Brian Franczak is a well-known dinosaur illustrator. His work has appeared in many magazines and books, including *Plant-Eating Dinosaurs* in the Prehistoric Life series, and on the PBS television series, "Nova."

Bob Walters is a dinosaur reconstruction artist whose paintings have been featured in numerous books and magazines. His work is displayed in The Smithsonian Institution and at The Academy of Natural Sciences, and has been seen on television specials for PBS and The Learning Channel.